A Manual for Translators of Mathematical Russian

A Manual for Translators of Mathematical Russian

S. H. Gould

Revised Edition, Edited by R. P. Boas

Published by the
American Mathematical Society
Providence, Rhode Island

1991 *Mathematics Subject Classification.*
Primary 01-00
Secondary 00A20

Library of Congress Cataloging-in-Publication Data

Gould, Sidney H. (Sydney Henry), 1909–
A manual for translators of mathematical Russian/S. H. Gould: edited by R. P. Boas.—2nd ed.
p. cm.
Includes index.
ISBN 0-8218-0172-4
1. Russian language—Translating into English. 2. English language—Translating into Russian. 3. Mathematics—Translating. 4. Russian language—Technical Russian. I. Boas, Ralph Philip. II. Title.
PG3498.G6 1991 91-20605
428′.029171—dc20 CIP

The paper used in this book is acid-free and falls within the guidelines established to ensure permanence and durability. ♾

This publication was typeset using $\mathcal{A}_{\mathcal{M}}\mathcal{S}$-TEX, the American Mathematical Society's TEX macro system.

1 2 3 4 5 6 7 8 9 10 95 94 93 92 91

Contents

Introduction

"The act of translating does not consist in carrying words across a no-man's land, but in answering the question: How would I say this if the notion occurred to me for the first time in my own language?"—Jacques Barzun

Good translators of mathematical Russian need three qualifications. Listed in order, from the most to the least important, these are, in a general way,

1. knowledge of mathematics;
2. knowledge of English;
3. knowledge of Russian.

Thus the most successful translators of mathematical Russian are competent mathematicians whose native language is English but whose knowledge of Russian is perhaps less than perfect. Each of the three qualifications may be fulfilled to a greater or lesser extent; and a deficiency in one respect may be balanced by greater competence in the others. Thus translators who have to cope with a somewhat unfamiliar branch of mathematics may be able to produce adequate translations if they know Russian extremely well and can express themselves well in English; those whose knowledge of English is less than perfect may be successful if they know the relevant field of mathematics very well in Russian. There are many reasons why translators with all three qualifications are scarce; consequently, editors who commission translations must often be satisfied with less than perfection.

The author of the first edition of this booklet, during four years' experience in editing translations, kept records of some unfortunate features, both grammatical and stylistic, that turned up in translations. The editor of the second edition did the same, for a longer time but under less intensive working conditions. The booklet was written, and revised, in the hope that, if we call attention to some of these solecisms, future translators will avoid them.

If you feel you would never make mistakes like those listed here, be assured that these examples were all taken from actual translations. If, on the other hand, you feel that some of the suggested improvements are rather trivial, please remember that unidiomatic word-order or syntax in translations, introduced under the influence of the original language, will make readers stop and ask themselves, perhaps subconsciously, "How is that again?" Whenever this happens, the translation is poorer than it ought to have been.

The main part of the booklet consists of typical examples. We first give the Russian text, then the faulty translation (in braces, which are sometimes omitted), then an acceptable translation, and usually some comments.

An unclassified list of examples would be of little value in itself. No two passages are exactly alike; translators must recognize, not a specific pitfall, but the species to which it belongs. We have therefore tried to formulate comments so that they will be of general interest.

Although the examples are taken from the mathematical literature, our advice should also be useful to physicists, chemists, engineers, and generally to anyone concerned with the translation of scientific Russian into English.

This manual is not intended for beginners, but for mathematicians who are already fairly well acquainted with Russian. It does not claim to be either complete or systematic, although it does deal with many common mistakes and misconceptions. It also does not discuss the problem of finding technically correct translations of individual words or phrases: for these, you should consult the *Russian-English Dictionary of the Mathematical Sciences* , published by the American Mathematical Society in 1961, and revised and enlarged in 1990. The English-Russian companion to the 1961 volume (**Англо-Русский словарь математических терминов** , Moscow, 1962), if available, can also be helpful.

Translators should, if possible, consult a specifically *mathematical* Russian-English dictionary. There are many general Russian-English dictionaries, but they are constructed primarily for students of Russian literature, history, etc. If translators pick definitions from general dictionaries, they frequently happen on "elegant" phrases that are rarely used in mathematical writing: for example, "in this manner" instead of "in this way" (**таким способом**); "stems from" instead of "follows from" or "is implied by" (**вытекает из**); "entail" instead of "imply" or "lead to" (**влечь за собой**).

It is best to avoid using abbreviations like *i.e.* , *e.g.* , and *cf.* Most present-day mathematicians have not studied Latin, do not know exactly what the abbreviations stand for, and consequently misuse them. The very common Russian abbreviation **т.е. (то есть)** stands for "that is" (*i.e.* in Latin). (The abbreviation *e.g.* is the Latin abbreviation for "for example.") The Russian **ср.** stands for **сравни** ("compare"; English *cf.*)

There are several important rules for translators:

1. The text you are translating probably meant something; therefore the translation must make sense.

2. If you do not understand what you are translating, don't fake it; if you do, you are likely to introduce errors. If all else fails, consult your editor for advice.

3. After completing a translation, read it against the Russian text. This is a check against both careless translations and unintentional omissions. Too many translators neglect this check, and editors don't have time to do it for them.

4. You must reread your translation after at least 24 hours. This is important, because while you are actually translating, you have the Russian text in front of you, and this advantage over the ultimate consumer takes time to overcome.

Incidentally, by no means all papers in Russian are written by literate native speakers of Russian. Sometimes you may have to correct the text that you are translating.

R. P. Boas

Part One
General Remarks on Grammar and Style

§1. Abstract Nouns

1. **Отсюда следует непустота множеств**

   [hence follows the nonemptiness of the sets]
   it follows that the sets are not empty

2. **ответить на вопрос о совпадении идеалов можно лишь при дальнейшем изучении многочленов**

   [the answer to the question concerning the coincidence of the ideals is possible only after further study of the polynomials]
   to determine whether these ideals coincide, we must make a further study of the polynomials

3. **это уравнение противоречит допущению о наличии в области G точек, не принадлежащих области D**

   [this equation contradicts the assumption concerning the presence in the domain G of points not belonging to D]
   This equation contradicts the assumption that G contains points that are not in D

Some of the numerous abstract nouns in present-day mathematical Russian (for example, **единственность** = uniqueness) can be translated by the corresponding English noun, but most of them need a different treatment. One objection, among many, to translating abstract nouns by abstract nouns is that in an uninflected language like English the result is often an unpleasant pile-up of prepositional phrases.

§2. Acknowledgments

1. **приношу глубокую благодарность О. А. Олейник за полезные советы и внимание к работе**

   [I express deep gratitude to O. A. Oleĭnik for his useful advices and attention to the work]
   I wish to thank O. A. Oleĭnik for her valuable advice and her interest in my work

2. **считаю своим долгом выразить благодарность профессору за постановку ряда задач**

   [I suppose it is my duty to express gratitude to my professor, who has posed a number of problems]
   I thank my professor for suggesting several problems

Russian acknowledgments are generally much more flowery than is customary in English. All polite expressions are notoriously dangerous; for example, **считаю своим долгом** cannot be rendered as "I suppose it is my duty."

§3. Adjectives (two or more)

1. **решение дифференциального линейного уравнения**

   [the solution of a differential linear equation ...]
   the solution of a linear differential equation ...

2. **пусть замкнутая жорданова спрямляемая кривая** ...

   [let the closed Jordan rectifiable curve ...]
   let the closed rectifiable Jordan curve ...

When two or more adjectives modify the same noun, the word-order is likely to be different in Russian and in English. The accepted word-order seems to be largely a matter of custom; translators need to be familiar with English usage if they are to avoid awkward phrases.

§4. Adverbs as syntax carriers

1. **оценка** (26) **устанавливается вполне аналогично оценке** (25)

   [the estimate (26) is set up completely analogously to the estimate (25)]
   the proof of (26) is just like that of (25)

2. **но тогда от этих** x, y **можно было бы перейти к еще меньшим параллельно переходу от** x_1, y_1 **к** x_2, y_2

   [but then from these x, y it would be possible to pass to still smaller parallelly to the passage from x_1, y_1 to x_2, y_2]
   but then we could proceed from the pair x, y to a still smaller pair in the same way as from x_1, y_1 to x_2, y_2

In Russian an adverb inherits more syntactical possibilities from its parent adjective than in English. In fact, syntactically speaking, the English adverb is a poor relation; we can say "this proof is similar to that proof," but not "this theorem is proved similarly to that theorem." The three adverbs that cause the most difficulty are **аналогично** = analogously, similarly; **подобно** = similarly, alike; and **параллельно** = in a parallel way (or better: correspondingly, along the same lines).

аналогично is usually weaker than "analogously"; and is better rendered by "similarly," unless the author is calling attention to an actual analog.

§5. Dangling gerunds (participles)

1. **следуя рассуждениям Н. Н. Баутина, теперь легко установить, что**

   [following the argument of N. N. Bautin, it is now easy to show that ...]
   ··· , we can now easily show that ...

2. **построение таких функций легко провести, несколько видоизменив метод, изложенный в (7)**

   [slightly changing the method explained in (7), the construction of such functions is easily carried out]
   such functions are easily constructed by slightly varying the method of (7)

In an inflected language like Russian, where a participle must agree in gender, number, and case with the noun it modifies, dangling participles are extremely rare. But Russian also has gerunds, for example, **следуя, читав, читавши, прочитав, прочитавши**, which look to an English reader like active participles. Since these forms, in contrast to true participles, are not inflected, they can look like dangling participles.

Some writers of English hold that modern grammar, like modern music or art, must break out of old forms if it is to achieve vivid expressiveness, whereas others, like the present writers, hold that a dangling participle is simply a solecism. We feel that a translation of a mathematical text is not a suitable proving ground for experiments in art. Even if we admit that a construction with a dangling particle is as perspicuous as any other, there remains the further requisite (referred to in the introduction as calling for a twenty-four hour reread): readers should be left free to think about the meaning of the text, not its grammatical structure; they should not be led subconsciously into expecting some grammatical feature and then stumbling into something else. Readers who see a participle will usually expect to see it supported later by a noun or pronoun. In example 2, the word "changing" brings to mind the set of mathematicians "we" (including at least the author and the reader) who have been active in the earlier part of the paragraph; the habits of a lifetime cause the readers to feel at least a slight tremor when the word "we" fails to turn up. A good style does not create such tremors.

Many idiomatic expressions in English involve what look like, and perhaps in some cases actually are, dangling participles; for example, "including," "assuming that," and "provided that." These expressions are unobjectionable, provided that they do not lead readers to expect a supporting noun, and they are becoming more numerous in English; but they are still a long way from including a phrase like "differentiating both sides, the equation becomes … ," and a mathematical text is not the place for a crusade on their behalf.

§6. Dash

1. **обозначим через** C **пополнение** S **по этой норме и через** C^***— соответствующее сопряженное пространство**

   let us denote by C the completion of S in this norm, and by C^* the corresponding dual space

2. **функции** L, M **удовлетворяют условиям, описанным в теореме 5, а** A, B, C**—условиям гладкости**

   [the functions L, M satisfy the conditions described in Theorem 5, but A, B, C—the conditions of smoothness]
   the functions L and M satisfy the conditions of Theorem 5, and A, B and C satisfy the smoothness conditions

3. **продолженные решения обладают свойством II—свойством равномерной непрерывности**

   the extended solutions have property II, namely uniform continuity

4. **при всех** n $f_n(x)$**—непрерывная**

   $f_n(x)$ is continuous for every n (Also see §16.)

It is often said that the Russian dash should be translated by "is" or "are," and in fact this is often a good translation (example 4); but in practice many translators simply replace the Russian dash by a dash in the translation; this is usually unsatisfactory. Examples 1–3 illustrate common possibilities: simply omit the dash; repeat the word represented; and translate the dash by "namely."

5. **пусть** $d(E)$ **период**

   let $d(E)$ be a period

Here we have an implicit dash. A similar example is **в каждой точке** E (at each point of E); here E was defined in a preceding sentence to be a set.

§7. Double negatives

1. **подгруппа** P_n **не содержится в централизаторе ни одного элемента группы** B...

   the subgroup P_n is not contained in the centralizer of any element in V

2. **пусть** Z_j^p – p**-мерные циклы, никакая линейная комбинация которых не гомологична нулю**

   let Z_j^p be p-dimensional cycles, no linear combination of which is homologous to zero

3. **если многообразие** M **таково, что** $A \neq B$ **ни для каких** S, **то многообразие** M **эквивалентно многообразию** N

   if the manifold M has the property that $A \neq B$ for every S, then M is equivalent to N

Double negatives often cause trouble, but the situation is really very simple. They are somewhat more common in Russian than in English (for an illustration, see under "can and may" in §5 of Part 3), but, with the single exception noted below, they are used in exactly the same way in both languages: that is, one negative cancels the other. Thus Turgenev's famous sentence **нельзя верить, чтобы такой язык не был дан великому народу** means "it is impossible to believe that such a language was not given to a great people." The single exception is that when **ни** is used in a sentence that contains the simple negative **не** (not), expressed or implied, the **ни** reinforces the meaning of **не**. In example 3 above, the **не** is implied by the inequality sign.

§8. Genitive after a negative verb

1. **если дуга не пересекает хорды**

   [if the arc does not intersect the chords]
   if the arc does not intersect the chord

If translators in a hurry forget that a negative verb is followed by a genitive (partitive in origin: that is, the arc does not intersect any part of the chord), they are likely to make various kinds of mistakes, for example, confusing singulars with plurals.

§9. Nouns followed by cases other than the genitive

1. **управление процессом передается этому элементу**

   [the control is transferred by the process to this element]
   the control of the process is transferred to this element

Most Russian verbs govern the accusative case. For example, **обобщать процесс** (to generalize the process); the corresponding noun is followed by the genitive case: **обобщение процесса** (generalization of the process). But many verbs govern some other case, for example, the instrumental. Thus we have **управлять процессом** (to control the process), and then the corresponding noun is also in the same case (translated by using "of" in English): thus we have **управление процессом** (control of the process).

§10. Order of subject and object, etc.

1. **более тонкое необходимое условие дает касательный пучок многообразия**

   [a sharper necessary condition provides the tangent bundle of the manifold]
   the tangent bundle of the manifold provides a sharper necessary condition

2. **неравенству** (5.1) **удовлетворяет конечное число пар**

   [inequality (5.1) satisfies a finite number of pairs]
   a finite number of pairs satisfy inequality (5.1)

3. **для этого при** $-\infty < k < \infty$ **и** $0 < \delta < \infty$ **введем множества** $E_n(k, \delta)\ldots$

   for this purpose we introduce sets $E_n(k, \delta)$ for $-\infty < k < \infty$ and $0 < \delta < \infty$

Whole books have been written about word-order in Russian. Ordinarily, Russian writers have greater freedom than writers in English, because the subject and object of a verb will be distinguished by their case endings. Thus "the dog sees the cat" can be written either **собака видит кошку** or **кошку видит собака**. In sentences like the preceding examples, however, only the context can indicate the correct English order. In such cases translators must be on the alert, since the English word-order may, or may not, be misleading.

§11. Participles as "slight" words

1. **прежде всего мы установим необходимость сформулированного нами условия**

   [first of all we shall establish the necessity of the condition formulated by us]
   we first establish the necessity of our condition

2. **в первом пункте проводимого нами доказательства**...

   [in the first point of the proof being carried out by us...]
   in the first section of this proof... (For the translation of **пункте** see Part Two, §18.)

3. **если исключить указанные возможности**

   [if we exclude the indicated possibilities]
   except for these possibilities

4. **по этой причине в разбираемом случае**

   [for this reason in the analyzed case]
   thus in the case in question

5. **удовлетворяет всем условиям леммы**

   satisfies the hypotheses of the lemma

One of the many effects of Russian's lack of definite and indefinite articles is the apparent superfluity in Russian of participles of all kinds, active and passive, present and past, preceding and following nouns. Very often the sole purpose of the Russian particle is to refer unambiguously to some preceding word, a task ideally performed by the English word "the."

Participles used in this way come under the heading of "slight words" (see §19, below) and are overtranslated by English participles. Often the best procedure is to omit the participle, translating it by the word "the," which will usually have to be used in any case.

If "the" does not seem to be enough, various other "slight" words or phrases are available, for example, "this," "these," "in question." If the participle is an honest one, even by the standards of a language with a definite article, it will usually come after the noun in English.

§12. Position of descriptive phrases

1. **для некоторых отличных от** z **точек** x, y

   [for certain different from z points x, y]
   for points x, y different from z

2. **из последовательности** $\{\psi_m(z)\}_0^\infty$ **можно выделить равномерно сходящуюся на любом замкнутом множестве из области** $|z| > \rho$ **подпоследовательность** $\{\psi_{m_s}(z)\}_{s=0}^\infty$

   we can extract from the sequence $\{\psi_m(z)\}_0^\infty$ a subsequence $\{\psi_{m_s}(z)\}_{s=0}^\infty$ that converges uniformly on every closed subset of the domain $|z| > \rho$

3. **последовательность возрастающих непрерывных функций** S

   [a sequence of increasing continuous functions S]
   a sequence S of increasing continuous functions

In English, a phrase that modifies a noun usually follows the noun, whereas Russian generally prefers to put the modifying phrase before the noun. A translator who is not aware of this construction is likely to translate example 3 as "a sequence of increasing continuous functions S," whereas the author intended to call the sequence S. Example 2 is even richer in possible mistranslations. Also compare §10, pages 7–8.

§13. Prepositional phrases

1. **поэтому интеграл, стоящий в правой части тождества** (5), **аналитически продолжается во всю область**

   [thus the integral standing on the right side of (5) is analytically continued to the whole domain]
   thus the integral on the right-hand side of (5) can be continued analytically to the whole domain

2. **характеристика, проходящая через точку** A, **не совпадает с характеристикой через точку** B

   [the characteristic passing through the point A does not coincide with the characteristic passing through the point B]
   the characteristic through A is not the same as the one through B

3. **буквой** M (**снабженной индексами**) **обозначаются положительные константы**

   [by the letter M (equipped with indices) are denoted positive constants]
   the letter M, with subscripts, will denote positive constants

In English, a prepositional phrase can be used equally well as an adjective or as an adverb. For example, in the sentence "the coat on the wall was made in Germany" the phrase "on the wall" is adjectival and the phrase "in Germany" is adverbial. In Russian a prepositional phrase is much more likely to be adverbial, usually as the result of inserting a participle; thus a Russian would say "the coat hanging on the wall . . . " Consequently it is wise, and at times almost mandatory, for the translator to leave some of the Russian participles untranslated. The three preceding examples are representative of countless others. (See also "participles as slight words," §11.)

§14. Repetition of words

1. **предположим, что** $F * G = H$, **где композиция** $*$ **определена по формуле** $F * G = RS$, **где** $R = TV$

   [we assume that $F * G = H$, where the operation $*$ is defined by $F * G = RS$, and $R = TV$]

   we assume that $F * G = H$, the operation $*$ being defined by $F * G = RS$, where $R = TV$

2. **вытекает существование факторизации** $N = N_1 N_2$, **где** N_1, N_2**— инвариантные подгруппы взаимно простых порядков, причем наибольшими** П-**делителями этих порядков являются соответственно** ν_1, ν_2

   [thus there exists a factorization $N = N_1 N_2$, where N_1 and N_2 are invariant subgroups of mutually prime order, where the largest П-factor of these orders is ν_1, ν_2, respectively]

   ... subgroups, with mutually prime orders, for which the largest П-factors are...

There are two complementary mistakes that involve repetition of words. In the first, the same Russian word is used twice in exactly the same setting, but the translator uses two different words in English, perhaps from a misguided attempt to improve the style. In the second, the Russian uses two different words, in different grammatical settings, but the translator, for any one of various insufficient reasons, uses the same English word twice. For example, it is objectionable to translate **область** G... **а область** H by "the domain G... but the region H... ;" on the other hand, it may be necessary to find different ways of translating **где** and **причем**. Sometimes this is difficult, but often either the **где** or the **причем** can be translated by a particle or a phrase.

§15. Sentence connection

1. **эти понятия, вообще говоря, разные. Так, например, в кольце матриц,** ...

   [these concepts are, in general, distinct. Thus, for example, in the ring of matrices , ...]

   Omit the "thus."

2. **Отсюда вытекает, что** ...

   [from this it follows that ...]

   Omit "from this."

The connection between successive sentences may be indicated either explicitly, by a word like "thus" or "but;" or else implicitly, that is, some words near the beginning of the second sentence remind the reader of words near the end of the first. Since the explicit method is commoner in Russian than in English, the translator has a golden opportunity for "translation by omission."

§16. Separation of symbols

1. **для** $n \in \Lambda_r\, m(H_n \cap \Omega_r) \leq \frac{1}{4}$

   when $n \in \Lambda_r$ we have $m(H_n \cap \Omega_r) \leq \frac{1}{4}$

2. **проведем через точку** $N\, n\, (n - 1)$**-мерных плоскостей** $y_i = C$, $i = 1, 2, \ldots, n, \ldots$

   [we draw through the point $N\, n\, (n-1)$-dimensional planes ...]
   through the point N we pass the n hyperplanes $y_i = C$ of dimension $n - 1 \ldots$

3. **пусть** $B(\lambda) \in A(L(H))$, $\sigma(B) \subset D$, $F(cD)$**—замкнутое множество и пусть** $C \backslash F$**—связное**

   let $B(\lambda) \in A(L(H))$, let $\sigma(B) \subset D$, let $F(cD)$ be a closed set and let $C \backslash F$ be connected

If two expressions, each consisting entirely of mathematical symbols, occur next to each other, it is natural for readers to assume that they form a single expression, or at least have parallel meanings. Otherwise they should be separated by words; commas are too likely to be overlooked. A sentence like "if P, Q" (meaning "if P, then Q") is marginally acceptable, but only because the "then" clause is very short. In example 1, above (in general, the difficulty is more insidious in longer sentences) readers will mentally begin "for $n \in \Lambda_r$ and $m(H_n \cap \Omega_r)$" and will then receive a mental jolt. Even in mathematics, where reading is slow, a style is bad if it forces readers to readjust their first impressions of the syntax.

In example 2, the trouble is even more pronounced; the three expressions N, n, and $(n - 1)$ should certainly be separated by words. As they stand, readers will almost certainly be confused by thinking that they are intended for a product.

In example 3, the commas do not provide adequate separation of the symbols; repetition of "let" overcomes this difficulty.

§17. Separation of verb and object

1. **выделим в прямом разложении** (19) **слагаемое** ...

   [we select in the direct decomposition (19) a summand]
   in the direct decomposition (19) we select a summand ...

2. **напоминаем, что мы не отличаем в записи эквивалентные матрицы**

   [we recall that we do not distinguish in our notation equivalent matrices]
   we recall that our notation does not distinguish equivalent matrices ...

3. **следует, что в течение процесса в этом слове находится точно один челнок**

   [it follows that in the course of the process in this word there is always exactly one shuttle]

it follows that in the course of the process there is always exactly one shuttle in this word

Separation of verb and object by a prepositional phrase is extremely common in Russian, but in English it is much rarer. The most important feature is the relative lengths of the phrase and the object: compare "he read with great interest the book he had received from his daughter," which is quite acceptable, with "he read with great interest the book," which is impossible.

The best solution is usually to put the prepositional phrase in front of the object, either at the beginning of the sentence (example 1), or in a "that" clause (example 2), immediately after the "that." Example 3 illustrates the amusing fact that, although English is much more likely than Russian to put a single prepositional phrase at the beginning, it cannot put two of them there (unless one of them merely modifies the other), whereas a Russian writer is quite free to do so.

§18. Singulars and plurals

1. **эти системы имеют как угодно медленно растущие множители**

   [these systems have arbitrarily slowly growing factors]
   each of these systems has a factor of arbitrarily slow growth

2. **локализуемость следует из постулатов Маклейна, о единичных сомножителях и ослабленного постулата Мальцева**

   [the localizability follows from the postulates of Mac Lane, on unit factors and the weakened postulate of Maltsev]
   the localizability follows from the Mac Lane postulate, the unit-factor postulate, and the weakened Maltsev postulate

3. **пусть p^a, $a > 0$,—любая и q^b, $b > 0$,—наивысшая степени различных или совпадающих простых чисел p и q, делящие соответственно g и h**

   [let p^a, $a > 0$, be an arbitrary and q^b, $b > 0$, be the highest powers of distinct or identical prime numbers p and q dividing g and h respectively]
   let p^a, $a > 0$, be an arbitrary power of p dividing g, and let q^b, $b > 0$, be the highest power of q that divides h, where p and q are primes, not necessarily distinct

It is generally true, at least in English, that singulars are clearer than plurals. Consider the sentence: "when wives are discussing problems with their husbands, they almost never know whether they understand them." Here everything is ambiguous, including whether a wife has more than one husband. How much clearer it is to say, for example, "when a wife is discussing a problem with her husband, he almost never knows whether she understands him."

Thus in a phrase like "these systems have factors of arbitrarily slow growth" (example 1), it is not clear whether each system has more than one factor; the phrase "the postulates of Mac Lane and Maltsev" leaves the readers uncertain whether Mac Lane and Maltsev share several postulates, or each has several of his own, or Mac Lane has one and Maltsev several, and so on. How much better to use the singular, with no fear of repetition of words: "the Mac Lane postulate and the Maltsev postulate."

The importance of translating in the singular is underlined by a feature of Russian syntax: In English we may say "he sent me a French book and a German book" or "he sent me a French and a German book", but not (as regularly in Russian) "he sent me a French and a German books" (**послал мне французскую и немецкую книги**). Without a clear understanding of this idiom, translators are likely to be helpless in the face of a sentence like example 3, where **любая** and **наивысшая** are singular adjectives modifying the plural noun **степени**.

§19. Slight words

1. **требуя лишь выполнения условий ортогональности матрицы** A**, что, как известно, всегда можно сделать**

   [requiring only fulfilment of the conditions of orthogonality of the matrix A, which, as is known, can always be done]
   requiring only, as is of course possible, that A is orthogonal

2. **таким образом, легко убедиться, пользуясь этими леммами, что**

   [in this manner it is easy to convince ourselves, using these lemmas that]
   consequently it is easy to see from these lemmas that...

3. **изложенным методом можно решать ряд практически важных задач**

   this method enables us to solve several problems of practical importance

4. **классы** A, B **тривиальны или нетривиальны одновременно**

   the classes A and B are either both trivial or both nontrivial

5. **при этом минимальный радиус этих кругов не меньше чем** r/n

   [in this connection, the minimal radius of these disks is not less than r/n]
   the minimal radius of the disks is at least r/n

6. **с помощью лемм 1 и 2 получаем следующую теорему**

   [with the help of Lemmas 1 and 2, we obtain the following theorem]
   we obtain the following theorem from Lemmas 1 and 2

7. **пользуясь приведенными выше результатами нетрудно доказать, что...**

   [using the results introduced above it is not difficult to prove that...]
   from these results, it is easy to show that...

8. **это—так называемая диаграмма Найквиста**

   this is the Nyquist diagram

Every language has many words or phrases that once were meaningful but have been used so often that readers are scarcely aware of them. If they are translated literally (and it is often hard to translate them in any other way), they are already overtranslated. A good example (number 1) is the phrase **как известно**, often translated by "as is known" (or perhaps better) by "as is well known." However, in many cases the author is referring to a mathematical fact that is indeed so well known that to call it so in English is absurd. Good alternatives are "as we know;" or "by a familiar theorem." We might also use some "slight" English word like "naturally," or "obviously," or even nothing at all.

Other examples are:

то, which is slighter than "therefore" or "then"
в связи с этим, slighter than "in connection with this result"
таким образом, usually much slighter than "in such a way"
ряд, often much slighter than "series"
убедиться, considerably slighter than "to convince oneself"
одновременно, slighter than "simultaneously"
в самом деле, slighter than "in actual fact"

In example 5 the slight force of **при этом**, often overtranslated, is probably carried sufficiently well by "the." In example 8, **так называемая** is often translated by "so-called," but this often has a pejorative connotation in English that is not present in Russian. It is usually better to use "known as," or to treat it as a slight word and leave it untranslated.

§20. Substitutes for indefinite articles

Strictly speaking, Russian does not have articles (compare §11), but from the translators' point of view one may say that Russian has equivalents for articles when they are needed. Many of the "slight" words in §19 can play the role of articles, and can advantageously be translated by articles.

1. **пусть f—произвольная аналитическая функция**

   [let f be any analytic function]
   let f be an analytic function

Here **произвольная** takes the place of the nonexistent "an" and is best so translated. Similarly, **некоторый, любой, какой-либо**, and **каждый** should often be translated by "a" rather than "a certain," "an arbitrary," "one or other," and "every." The very common title **Об одной теореме**... seems much more normal in English as "On a theorem ... " than as "On one theorem ... " or even "On a certain theorem."

2. **определим теперь важную операцию**

   we now define (the *or* a?) fundamental operation

Here the choice between "a" and "the" depends on the context: is there only a single operation, or are there many operations?

3. **Об одном характеристическом свойстве функций вполне регулярного роста внутри угла**

   On a characteristic property of functions of completely regular growth in an angle

Here the quoted phrase was the title of the paper; since no particular angle was specified, "an" is appropriate.

§21. Symbols as syntax carriers

1. **для некоторого** $0 < k < M$

   "for some $0 < k < M$" is used by some writers in English, but a better style is "for some k, $0 < k < M$"

2. **так как рассматриваемые** $n < 2p^2 - 7p + 7$, **то** ...

   [since the considered $n < 2p^2 - 7p + 7$, then ...]
   since the values of n in question are all less than $2p^2 - 7p + 7$, it follows that ...

3. **оператор, заданный на** $u \in C^2$, **удовлетворяющих** $u(0) = u(1) = 0$ ...

   [an operator defined on $u \in C^2$ satisfying $u(0) = u(1) = 0$ ...]
   an operator defined for functions u that satisfy $u \in C^2$ and $u(0) = u(1) = 0$ ...

For several reasons, Russian writers can assign much greater syntactical activity to mathematical symbols than is possible in English. The chief reason is that in Russian a modifying participle can indicate the case of a noun that is represented by a collection of symbols. In example 2, the participle **рассматриваемые** is in the nominative plural and therefore must modify the plural noun implied by the symbol n; this, in turn, implies that the symbol $<$ must stand for the main verb of the sentence. Choreography of this sort is scarcely possible in English.

In example 3, the participle **удовлетворяющих** is in the prepositional case and therefore cannot modify anything in the preceding part of the sentence except the symbol u; here again some readjustment is required in English.

§22. Symbols in the genitive case

1. **элементы пространства** W **различны**

   [the elements of the space W are distinct]
   the elements of W are distinct

2. **получим пространство** W; **свойства элементов** W **очевидны**

   we obtain the space W; the properties of the elements of W are obvious

3. **из нормальности типа** $f(z)$ **следует, что** ...

   [from the normality of the type of $f(z)$ ...]
   since $f(z)$ is of normal type ...

4. **каждая непустая порция** $[a, b]$ **имеет положительную меру**

   every nonempty part of $[a, b]$ has positive measure [*or*: every nonempty part $[a, b]$ has positive measure?]

5. **для того, чтобы некоторая подполугруппа** S **была изоморфна**

   for some subsemigroup S to be isomorphic [*or*: for some subsemigroup of S to be isomorphic?]

6. **группа** G **действует локально транзитивно в точке** $m \in M$

   a [*or* the?] group G acts locally transitively at the [*or* a?] point m

Mathematical symbols are, of course, not inflected, so that, in example 1, Russian cannot say "the elements of W are distinct," but has to say "the elements of the space W are distinct." The word **пространства** merely indicates that the symbol W is in the genitive case; the English "of" performs the same task very well. However, some Russian writers disregard the rule, as in example 3, apparently on the principle that the syntax will be clear from the context; the result can be very troublesome. Translators often have to read a sizeable amount of text, with great care, in order to decide what case a symbol has and consequently what the Russian writer intended.

In example 2, W is in turn accusative and genitive. In example 4, it is hard to say at first glance whether $[a, b]$ is genitive or nominative. In example 5, a translator might have to read the whole article to decide whether S is nominative or genitive. In example 6, a decision can be reached from the sentence that preceded the quotation: it defined G as a transformation group on a manifold M, so that a reasonable translation might be "G acts locally transitively at $m \in M$."

In cases like these, it is quite impossible for a translator to take refuge, as is so often done with other ambiguous passages, in a mere reproduction in English of the ambiguity.

§23. Translation of handwriting

Translators sometimes have to deal with Russian articles that have been reproduced from typewritten manuscripts with the mathematical symbols inserted by hand; or even with completely handwritten manuscripts. It may be hard to identify handwritten letters, particularly when algebraists use German letters, and write them in a typewritten manuscript in old-fashioned script, which has several variant forms. These letters will eventually be printed as old-fashioned German letters (Fraktur), so translators should treat them accordingly. The commonest convention is to type the corresponding ordinary (Latin) letters and identify them as German letters. It is usually better not to attempt to copy the shape of the German letter in the translation.

Also, in a Russian typewritten manuscript our ordinary italic letters will have been inserted by hand, possibly printed, possibly written in either Western script or Russian script. These should usually be identified with ordinary italics. (The present writers have never encountered Russian script in a typeset Russian mathematical paper.)

§24. Translation of mathematical formulas

Russian abbreviations for trigonometric and hyperbolic functions usually differ from those that are customary in English text. Translators should not merely copy formulas verbatim, or strip them into translations, but translate them.

The usual Russian symbols are as follows.

Trigonometric functions: sin and cos are the same in both languages. Russian uses tg for tangent and ctg for cotangent.

Inverse trigonometric functions: These are indicated, as in English, by a superscript -1 or else by the prefix arc or Arc.

Hyperbolic functions: $\mathrm{sh} = \sinh$, $\mathrm{ch} = \cosh$, $\mathrm{th} = \tanh$, $\mathrm{cth} = \coth$. Some Russian authors use the German convention that represents sinh by Sin in Fraktur.

Inverse hyperbolic functions: Their names are **ареасинус**, etc. The corresponding symbols are Arsh, etc.; or the name of the function, with exponent -1.

$y = \overline{1, n}$ means $y = 1, \dots, n$.

Superscripts **в** and **н**, which stand for **верхний** (upper) and **нижний** (lower) should be replaced by u and l; similarly for other subscripts and superscripts.

When a formula is too long to fit on a single line, the Russian convention is that when the formula is broken at the end of a line, an operator ($+$, $-$, $\times$, $=$, $\div$, etc.) is repeated at the beginning of the next line. In English, it is customary to use the symbol only at the beginning of the next line. You will need to take account of this convention if you are stripping formulas into your translation.

§25. Verbal adjectives (passive)

1. **оператор** A **называется ядерным, если он представим в виде...**

   [the operator A is called a nuclear operator if we represent it in the form...]

   ... is called an operator of trace class if it can be represented in the form...

2. **для этих уравнений применим принцип максимума**

   [for these equations we shall apply the maximum principle]

   the maximum principle is applicable to these equations

The present passive particle of an imperfective verb, for example **представлять** (to represent), is

представляемый, представляемая, представляемое,

and for a perfective verb, for example **представить**, the corresponding formation occurs very frequently in mathematical literature in the shortened predicative forms

представим, представима, представимо.

Nevertheless, the grammars assert that perfective verbs have no present passive participles.

The reason for this assertion is as follows. In the first place the perfective aspect is considered to have no present tense of any kind; for example, the perfective **представлю** is a future (I shall represent), so that **представим** may be called a future passive participle, with some such meaning as "about to be represented." However, the grammars do not call it a participle at all, but rather a "verbal adjective," since its actual meaning is almost never "about to be represented," but "such as can be represented," or "such as ought to be represented."

With this situation clearly in mind, translators will be able to avoid the common mistake of confusing the verbal adjective **представи́м** (representable) with the first person plural indicative **предста́вим** (we shall represent).

In example 1, the translation "if we represent it... " is simply a blunder, since **он** cannot be accusative. In example 2, it is probable that a Russian reader would automatically take **применим** as a verbal adjective, even though grammatically it could be a first person indicative.

Part Two
Russian Words and Phrases

§1. **большой**

1. **о представлении больших чисел (бóльших чисел)**

   on the representation of large numbers (larger numbers)

2. **с бóльшим числом**

   with a greater number

3. **с показателем на единицу большим, чем**...

   with exponent greater by unity than...

The comparative degree (greater) of the adjective **большой** (great) usually occurs in the predicate in the uninflected form **больше**. But an inflected form is sometimes used, especially in the genitive and instrumental cases, for example, **больших** and **большим**. In speaking, the accent on these comparative forms falls on the first syllable, in contrast to the positive forms, which are otherwise indistinguishable from them. Thus Russian texts customarily print an accent on the comparative forms except where it is clear from the context (example 3) that the comparative is intended.

§2. **в дальнейшем**

For some unknown reason, translators are apt to render this phrase by "in the sequel," which makes sense only if there is actually a following part. More idiomatic translations are "in what follows," or "later," or "in the next section," etc.

§3. **должно**

1. **Нетрудно убедиться, что найдется такое натуральное число** i. **Но тогда** C_{i+1} **должно содержать точку** x.

   [It is easy to show that there exists such a positive integer i. But then C_{i+1} should contain the point x.]

   ... But then C_{i+1} must contain the point x.

Although there are many places where **должно** ought to be translated by "should," it sometimes needs to be translated by "must." In the example, the translation "then C_{i+1} should contain the point x" implies, incorrectly, that it ought to but in fact does not.

§4. **если же**

1. **если** k **нечетно, то и** $k+2m$ **нечетно, если же** k **четно, то** γ **не квадрат**

   if k is odd, $k+2m$ is also odd; but if k is even, γ is not a square

2. **предположим сначала, что** $a \geq 0$; **имеем** ... ; **если же** $a < 0$, **то** ...

   we first assume that $a \geq 0$, when we have ... ; but if $a < 0$, then ...

In a mathematical passage the best translation of **если же** is always "but if;" at least, the writers have never found a counterexample. The particle **же** can be translated in many other ways in other contexts, but not after **если**.

§5. **замена**

1. **с заменой** k **на** m

   [with the substitution of k for m]
   with k replaced by m;   *or:*   with the substitution of m for k

Translators sometimes confuse the two words "substitution" and "replacement." The English word "change" is ambiguous, as in Ben Jonson's song "... But might I of Jove's nectar sup, I would not change for thine;" but there is no ambiguity about the corresponding Russian word **замена**. Thus, **замена синей книги черной** (exchange of a blue book for a black book) can only mean that the speaker originally had a blue book and now has a black book. A translator who hastily writes "substitution of a blue book for a black one" has made the kind of elementary blunder that makes nonsense out of any mathematical passage.

§6. **заметим**

The translation "we note" tends to be overworked. The parent verb **заметить** can mean "notice, observe, remark," and also "note" in the sense of "make a note of." Unless the Russian author actually wants the reader to make a note of something, one of the other possible translations is usually more satisfactory.

§7. **и**

1. **это я и сделал**

   [I also did this]
   this is what I did

2. **получим** $\sup k \geq 1$, **что и приводит к 5.26**

   [we obtain that $\sup k \geq 1$, which also leads to 5.26]
   thus $\sup k \geq 1$, as needed for 5.26

3. **можно брать полиномы того же вида, что и** ϕ

[we may take polynomials of the same form as ϕ also is]
... of the same form as ϕ

Hilaire Belloc wrote an essay on the subtleties of "and." The Russian word **и** is equally subtle, but the subtleties are different, so that the translator must sometimes reject the automatic translation "and" or "also." In example 2, "also" is quite wrong, since it implies that some other result also leads to 5.26; the meaning is something like "exactly" or "specifically," but either of these words is an overtranslation. The standard procedure is that used in example 1, but often it cannot be applied. In example 3, nothing seems feasible except to disregard the **и** altogether.

§8. **или, что то же**

This means literally "or, what is the same," but this is awkward, and the same idea is expressed more idiomatically by "or equivalently" or "or alternatively," depending on the context.

§9. **именно**

The usual dictionary translation is "namely," which English dictionaries define by "that is to say." Either of these terms has a somewhat pedantic flavor, and it may be preferable to select an alternative from among such terms as *exactly*, *just*, *precisely*, *specifically*, or *to be exact*.

§10. **каково бы ни было**

1. **каково бы ни было** $a > 1$ **, можно построить ...**

   [whatever $a > 1$ may be, it is possible to construct ...]
   for arbitrary $a > 1$, we can construct ...

If the phrase **каково бы ни было**, or any of its variants, is translated in any way other than "for arbitrary ... " or "for every ... ," the result will probably seem clumsy.

§11. **ли**

1. **интересно, сохраняют ли метризуемость произвольные регулярные отображения**

   [it is interesting whether metrizability is preserved under arbitrary regular mappings]
   it is an interesting question whether ...

The range of words that can introduce an indirect question is wider in Russian than in English, as in this example.

§12. **называть**

1. **тором** T^k **называем прямое произведение** k **окружностей**

   [a torus T^k is called the direct product of k circles]
   the Cartesian product of k circles is called a torus

2. **если** G **ранга** 1, **то** $B(G)$ **назовем открытое множество, которое**...

   [if G is of rank 1, we shall call $B(G)$ an open set which...]
   if G is of rank 1, then $B(G)$ will denote an open set which...

3. ... **главной частью оператора** L **назовем оператор** L_0... **и число** $2p$ **назовем обобщенным порядком уравнения**

   [... the principal part of the operator L we shall call the operator L_0... and the number $2p$ we shall call the generalized order of the equation]
   ...L_0 is called the principal part of L, and $2p$ is the generalized order of the equation

4. **будем называть** K**-циклом** Δ**-цикл, являющийся** K**-множеством**

   [we shall call a K-cycle a Δ-cycle if it is a K-set]
   A Δ-cycle that is also a K-set will be called a K-cycle

5. **будем называть** k**-мерные интервалы интервалами направления**

   [we will call k-dimensional intervals intervals of direction]
   the k-dimensional intervals will be called direction intervals

Consider the definition "a *torus* T^k is the Cartesian product of (circular) circumferences." Here, as is usual in mathematical writing in English, the concept to be defined (the definiendum) is italicized, and the concept in terms of which it is defined (the definiens) is left unitalicized. In the corresponding Russian (see examples 1 and 2) the definiendum **тором** (or in example 2 the symbol $B(G)$) is in the instrumental case (usually not italicized) and the definiens is in the accusative after the verb **называем** (or in the nominative with the reflexive verb **называется**). But many variations are possible in both languages, with consequent difficulties of translation.

In Russian, all six permutations of the word-order **тором называем произведение** actually occur (note the variation in word-order in example 3), and translators must take care not to interchange the roles of definiens and definiendum, since the resulting confusion (as in example 4) will cause acute discomfort in the reader's mind.

In English, it is usually desirable to let the verb separate the definiens from the definiendum. A juxtaposition like **интервалы интервалами** (example 5) is acceptable in Russian (and in fact is quite common), but in English "intervals intervals" is highly objectionable.

§13. некоторый

1. **напомним, что о некотором ядре K мы говорим, что оно допускает биортогонализацию в H, если...**

   [we recall that concerning some kernel K we say that it admits a biorthogonalization in H if...]
   we recall that a kernel K is said to admit...

2. **где p—некоторый параметр**

   [where p is some parameter]
   where p is a parameter

3. **при этом гомоморфном отображении соотношения между высказываниями перейдут в некоторые соотношения между переменными**

   under this homomorphism the relations among the propositions will be mapped onto relations among the variables

The (correct) belief that Russian has no indefinite article has been elevated by some translators into the (false) doctrine that **некоторый** should never be translated by "a," nor its plural **некоторые** by the plural of "a," namely nothing at all. The three examples above are given in the hope of undermining this doctrine.

§14. обозначить

1. **обозначим ее Γ**

   let us call it Γ

2. **обозначим еще $f = \max T(x)$**

   let us also set $f = \max T(x)$

3. **обозначим через S оператор ортогонального проектирования... обозначим далее $P = E - S$**

   we denote by S the operator of orthogonal projection... ; and we also set $P = E - S$

The one Russian verb **обозначить** can do the work of several English verbs: "designate," "denote," "call," "set." Confusion among them can lead to very unidiomatic expressions.

§15. общий

1. **два элемента пространства имеют общее расположение, если изображающие их плоскости имеют $(m+1)$ общих перпендикуляров**

   [two elements of the space are in general position if their representative planes have $(m+1)$ general perpendiculars]
   ... have $(m+1)$ perpendiculars in common

The Russian word **общий** can mean either "general" or "common;" occasionally it may be difficult to decide between them.

§16. по

1. **разложение по собственным функциям**

   an expansion in eigenfunctions

2. **разложение функций в евклидовом пространстве по функциям** ϕ_k

   [expansion of functions in Euclidean space according to functions ϕ_k]
   expansion of a function in Euclidean space in a ϕ series [*or*, a series of the functions ϕ_k]

3. **по известной теореме**

   by a familiar theorem

4. **по определению индуктивного предела**

   by the definition of an inductive limit

5. **операторы сравниваются по методу Л. Хермандера**

   the operators are compared by Hörmander's method

In many cases "according to" is not an acceptable translation for the ubiquitous Russian **по**. In fact, it is sometimes hard to find a good translation. Even the common phrase "expansion in a series of functions" (see example 2) can cause trouble.

§17. пока, пока не

1. **челнок будет повторять этот цикл, пока не сможет выполнить** (2); **пока сможет выполнить** (2)

   the shuttle will repeat this cycle until it can satisfy (2); until it cannot satisfy (2)

The Russian word **пока** means "as long as" or "while," and in spite of an apparently widespread opinion it does not mean "until." Consequently **пока не** (often with intervening words) means "as long as ... not," which is usually best translated by "until"; thus "they waited as long as I did not arrive" is much less idiomatic in English than "they waited until I arrived." Therefore **пока** alone, as the negation of **пока не**, can be translated not only by "as long as," but also by "until ... not." Compare "we waited as long as we could" and "we waited until we could not wait any longer." If the **не** is widely separated from the **пока**, translators who are in a hurry sometimes translate **пока ... не** by "until ... not," with consequent shipwreck of the sense.

§18. **пункт**

1. **доказательство разобьем на несколько пунктов**

   we present the proof under several headings

The Russian word **пункт** means an “item,” “heading,” or “subsection,” usually numbered. It also occasionally means “point,” as in the phrase **исходный пункт** (“starting point”). The word **параграф** means “section,” and the Russian for “paragraph” is **абзац**.

§19. **работа**

1. **следует из общих теорем работ** [17], [18]

   [it follows from the general theorems of the works [17], [18]]
   it follows from the general theorems of [17] and [18]

2. **в работе**

   [in the work]
   in this article [or paper, or book, etc.]

The frequently encountered “in the present article” is unnecessarily stilted. When **работа** refers to a definite book or article, the translation “work” is often unidiomatic; **работа** should then be translated by “book” or “article,” depending on which it is. However, **работа** can often simply be omitted.

§20. **следует из того факта**

1. **следует из того факта, что верен указанный результат**

   [it follows from the fact that the indicated result is correct]
   it follows because the result is correct

The phrase “from the fact that” is a common piece of jargon much used by authors and translators who are in a hurry. It should be avoided whenever possible.

§21. **тот или иной**

1. **этот пример наводит на мысль изучать распределение тех или иных функционалов**

   [this example suggests that we should study the distribution of these or other functionals]
   ... of the various functionals

The Russian phrase **тот или иной** does not mean “this or another” but rather “one or another,” “some or other,” and can usually be translated by “various.”

§22. **убеждаемся**

Although **убеждаться** means "verify, make sure of," it is often better not to translate **убеждаемся** by "one can convince one's self," but simply by "we see" or "we can see."

§23. **уже**

1. **на плоскости это утверждение справедливо. Уже на поверхности прямого кругового конуса это может быть не так**

   On the plane this statement is correct. However, even on the surface of a right circular cone there may be exceptions

2. **Остальная часть доказательства проводится уже более или менее просто**

   The rest of the proof is now more or less straightforward

3. **все эти подгруппы сопряжены с F уже с помощью элемента из G**

   each of these subgroups is seen to be conjugate to F under transformation by an element of G

4. **результат уже не получается**

   the result can no longer be obtained

In Russian, as in German and elsewhere, the word for "already" often causes trouble, since care must be taken not to overtranslate it. The first two examples show that words like "even" or "now" are sometimes available. However, the third example is more difficult: the meaning is that even the subgroup G is large enough to provide elements that will transform the subgroups into F. Unfortunately the English word "already" is not idiomatic in this setting, and it is hard to find another word or phrase that is short enough not to be clumsy. Admittedly, something is lost by omitting **уже**. In a negative sentence (example 4), "already... not" usually becomes "no longer" in English.

Part Three
English Words and Phrases

§1. "a" *or* "the"?

1. **пусть** F**—Фраттини-подгруппа группы** G

   [let F be a Frattini subgroup of G]
   let F be the Frattini subgroup of G

2. **совокупность всех элементов коммутативной группы** G**, порядок которых есть степень** p**, называется примарной компонентой, или** p**-компонентой, группы** G

   the set of all those elements of an Abelian group G whose orders are powers of a given prime p is called a primary component, or the p-component, of G

The absence of definite and indefinite articles in Russian often makes it hard to decide whether a word like **подгруппа** should be translated by "the subgroup," "a subgroup," or simply "subgroup." Consider example 1. If G has exactly one Frattini subgroup (as in fact it certainly does), we must write "the," but if more than one, then "a." A translator who is not sure which is the case, and does not have time to find out, may try something like "let G be a group and H a Frattini subgroup;" but the subterfuge is seldom successful.

§2. and

1. **машина обладает некоторой клеткой, конечным числом состояний и всегда находится в одном из них**

   [the machine has a cell, a certain number of states and is always in one of them]
   ... a cell and a certain number of states ...

2. **все критерии полиномов наилучшего приближения: Чебышева, Колмогорова, Зуховицкого и Стечкина**

   [all the criteria for polynomials of best approximation: namely those of Chebyshev, Kolmogorov, Zukhovitskiĭ and Stechkin]
   Insert "and" before Zukhovitskiĭ.

3. **все элементы являются пределами произведений элементов вида** $\psi^{-1}\phi\psi$, $\psi \in F$, $\lambda \cdot \mu$**—непрерывный гомоморфизм**

   [all the elements are limits of products of elements of the form $\psi^{-1}\phi\psi$, $\psi \in F$, $\lambda \cdot \mu$ is a continuous homomorphism]
   Insert "and" between $\psi \in F$ and $\lambda \cdot \mu$

One of the commonest reasons for unsatisfactory translation of Russian mathematics is translators' failure to remember that Russian often omits "and" where it is necessary in English. For example, the usual (though not invariable) Russian way of saying "let us construct a triangle, a circle, and a square" is "let us construct a triangle, a circle, a square."

In example 2, the Russian text means that Chebyshev and Kolmogorov have one criterion apiece, and Zukhovitskiĭ and Stechkin have one between them; if the latter two authors had one apiece, the **и** in the Russian would likely be replaced by a comma.

In a sentence like example 3, where the required "and" comes between mathematical symbols (or just before or just after a symbol), translators are particularly likely to omit it, with harsh results.

§3. anomalous use of "if"

1. **если в известном способе Литтльвуда уже в первом шаге навязывается избыточный множитель, то неравенство** (3) **этого недостатка не имеет**

   [if in the well-known Littlewood method already in the first term there intrudes a redundant factor, then inequality (3) does not have this disadvantage]

   in the well-known Littlewood method, a redundant factor is introduced even in the first term, but inequality (3) does not have this disadvantage

2. **если числа** l_k **естественно считать равномерно ограниченными, то для** m_k **дело обстоит иначе**

   [if the numbers l_k may naturally be taken to be uniformly bounded, then the situation is different for the m_k]

   although it is natural to make the assumption that the l_k are bounded, the situation is different for the m_k

3. **если классический математический анализ оперирует с "хорошими" функциями, то теория функций действительного переменного изучает значительно более общие классы функций**

   if classical analysis operates essentially with "nice" functions, the theory of functions of a real variable investigates considerably wider classes of functions

Certain languages, including English and Russian, have an idiomatic use of the word "if," in which the phrase "if P, Q" does not mean "if P then Q" but "P and also Q," with the implication that, in view of the known truth of P, the assertion Q is slightly surprising. Thus Chekhov's description of the island of Sakhalin:

если в Александровском округе климат морской, то в Тымовском он континентальный

may be translated

"if the climate is maritime at Aleksandrovsk, it is continental at Tymovsk."

Compare an English sentence like "if his parents were poor, they were honest."

However, if Russian mathematicians use this construction, it is extremely rare in mathematical English, and translators should avoid it altogether. If they insist on using it, they must remember that the Russian text may lead them astray. For example, if **то** is translated by "then," the whole passage is reduced to nonsense.

§4. both

1. **но обе функции, входящие в** (4), **непрерывны, и потому обе части** (4) **равны**

   [but both the functions in (4) are continuous, and thus both sides of (4) are equal]
   but the functions in (4) are both continuous, and thus the two sides of (4) are equal

It is a solecism in English to use the word "both," instead of "the two," in a statement which, usually because of the presence of some word like "together" or "equal," becomes nonsensical when applied to one person or thing. Thus "the numbers are both large" but "the two numbers are equal."

§5. "can" *and* "may"

1. **следует, что решения могут иметь в** D **лишь конечное число нулей**

   [it follows that the solution may have only a finite number of zeros in D]
   ... can have only a finite number ... ; *or* ... cannot have more than a finite number of ...

2. **если** $n = 1$, **то мы можем этого требования не налагать**

   [if $n = 1$, we may not impose this condition]
   if $n = 1$, we do not need to impose this condition; *or* we may dispense with this condition

3. **если** $n = 1$, **то мы не можем налагать этого требования**

   [if $n = 1$, we may not impose this condition]
   if $n = 1$, we cannot impose this condition

4. **если** $n = 1$, **то мы не можем не налагать этого требования**

   [if $n = 1$, then we may not fail to impose this condition]
   if $n = 1$, this condition cannot be discarded

It is true that in English "may" is sometimes more elegant than "can;" for example, "we may assume that n is prime." But "can" is much safer, especially with such words as "not" and "only." The phrase "may not" is ambiguous in English. For example, "they may not appear" means either "perhaps they will not appear" or "they are not allowed to appear." In Russian, the corresponding ambiguity is avoided by the word-order (examples 2 and 3). Double negatives, as in example 4, are commoner in Russian than in English.

§6. if and only if

1. **R считается реализацией формулы $T = S$ в том и только в том случае, когда ...**

   [R is considered to be a realization of the formula $T = S$ in the case and only in the case when ...]
   R is said to be a realization of $T = S$ if and only if ...

Russian has many variants for "if and only if" besides the one used in example 1, but the phrase does not vary in English.

§7. it

1. **неравенство $f(x) \leq 0$ является следствием системы $f_i \leq 0$; т. е. ему удовлетворяют все ее решения**

   [the inequality $f(x) \leq 0$ is a consequence of the system $f_j(x) \leq 0$; i.e. all its solutions satisfy it]
   ... that is, all the solutions of the latter satisfy the former

2. **из определения функции ϕ непосредственно вытекает, что она удовлетворяет условиям**

   [from the definition of the function ϕ it follows immediately that it satisfies the conditions]
   from the definition, it follows at once that ϕ satisfies the conditions

3. **нетрудно проверить, что можно взять ...**

   [it is easy to check that it is possible to take ...]
   it is easy to verify that we may take ...

4. **рассмотрим функцию $g(z)$, обладающую следующим свойством: точка $z = 0$ является ее существенно особой точкой**

   [we consider a function $g(z)$ with the property that $z = 0$ is its essentially singular point]
   ... $z = 0$ is an essential singular point of $g(z)$

Note the profusion in these examples of Russian words for "it": **ему, ее, она**; in English, some of them have to be translated by the corresponding nouns. In example 1, the gender of **ему** and **ее** cannot be reproduced in English.

The word "it" can also cause trouble in many other ways. For example, a representative subject immediately followed by "it" as an ordinary pronoun (example 2), or by another representative subject (example 3), makes an unpleasant impression. Also, the word "its" is tricky. Thus "its singular point" necessarily implies in English that the function has only one such point, whereas the corresponding Russian phrase **ее особая точка** (example 4) leaves this question open.

§8. -ly words

1. **совершенно аналогично доказывается**

   [entirely analogously it is proved that]
   in complete analogy with this result, we may prove that...

2. **эти R_M-полупростые кольца автоматически являются сильно R_M-полупростыми**

   [these R_M-semisimple rings are automatically strongly R_M-semisimple]
   it follows automatically that these R_M-semisimple rings are strongly R_M-semisimple

Two successive -ly words, one of them modifying the other, are objectionable in English; consequently considerable trouble should be taken to avoid them. The corresponding juxtaposition of adverbs is fairly common in Russian.

In example 1, the word **аналогично** is appropriately translated as "analogously." In most contexts, it has the weaker sense of "similarly."

§9. respectively

1. **компактная группа Ли тогда и только тогда нильпотентна (разрешима), если ее конечные подгруппы нильпотентны (соответственно разрешимы)**

   a compact Lie group is nilpotent (solvable) if and only if all its finite subgroups are nilpotent (solvable)

The example illustrates a common method of using parentheses to avoid repetition of phrases. In English, "respectively" is seldom inserted in the second parenthesis, and in general the word "respectively" is used far less often in English than in Russian.

§10. since... , then...

1. **так как функции w_i являются решениями этого уравнения, то разность любых из них можно рассматривать как решение некоторого линейного уравнения**

   [since the functions w_i are solutions of this equation, then the difference of any two of them may be regarded as the solution of a certain linear differential equation]
   Omit the "then."

2. **поскольку $z(x, c)$ есть монотонно возрастающая функция по c при фиксированном значении $x \in (0, l]$, то $\mu(x) \equiv \nu(x)$.**

   [since $z(x, c)$ is a monotonely increasing function with respect to c for a fixed value of $x \in (0, l]$, $\mu(x) \equiv \nu(x)$]
   since, for fixed $x \in (0, l]$, the function $z(x, c)$ is monotone increasing in c, it follows that $\mu(x) \equiv \nu(x)$

The combination "since... , then ... " (**так как... , то...**) is extremely common in mathematical Russian but totally inadmissible in English. When a signpost is needed in English (example 2) to show where the principal clause begins, the best one is usually "it follows that." If this phrase seems too ponderous, translators can fall back on the stereotyped "we have."

§11. such that

1. **возьмем такую точку на прямой, чтобы...**

   [we take such a point on the line that...]
   we take a point on the line such that...

2. **такая последовательность сечений, для которой...**

   a sequence of sections such that ...

3. **выберем во вмещающем пространстве базис так, чтобы...**

   we select a basis in the underlying space such that...

4. **число** n **выберем так, что** $n^2 > 100$

   we choose the number n so that $n^2 > 100$

5. **выберем такое** n, **чтобы** $n - 1$ **чтобы на** 7

   we choose a number n such that $n - 1$ is divisible by 7

Examples 1–3, and example 4, §7, give various possibilities in Russian for the common mathematical phrase "such that." The original author felt strongly that there is no corresponding freedom of choice in English; the editor of this revision, however, feels that alternatives like "with the property that" and "for which" can offer an opportunity for avoiding tedious repetition in translations.

Translators should be aware of the difference between **такой** and **так**: the first means "such" and the second means "so." In some cases these two words are interchangeable, but in examples 4 and 5 they are not.

§12. "the" *instead of* "this" *or* "that"

1. **носитель преобразования Фурье данной функции называется спектром этой функции**

   the support of the Fourier transform of a given function is called the spectrum of the function

2. **связь рациональных отображений многообразия в проективное пространство с классами дивизоров на этом многообразии**

   the connection between the rational mappings of a variety into a projective space and the divisor classes on the variety

Translators from Russian to English should watch for opportunities to replace "this" or "that" by "the." A good working rule is to use "the" for **этот** in the (quite common) places where the only purpose of **этот** is to refer unemphatically to a preceding word (compare "Participles as 'slight' words" in Part One, §11).

§13. whose

1. **пусть** X**—такое пространство, из всякого открытого покрытия которого мощности** r **можно выделить конечное покрытие**

   [let X be a space from every open cover of cardinality r of which a finite cover can be selected]
   let X be a space such that from every open cover of cardinality r we can select a finite cover

2. **сильно позитивный оператор, область определения которого не зависит от** t

   [a strongly positive operator the domain of definition of which does not depend on t]
   ... whose domain does not depend on t

3. **мы получим сумму дробей, числители которых ограничены, а знаменатели имеют следующий вид**

   [we obtain a sum of fractions the numerators of which are bounded and the denominators of which have the following form]
   a sum of fractions with bounded numerators and with denominators of the following form
   or: whose numerators are bounded and whose denominators have the following form

4. **разорвем компоненту** $\alpha - \beta$ **на несколько других, разным из которых и принадлежат** A **и** B

   [we break the component $\alpha - \beta$ into several others, to different ones of which A and B belong]
   we break the component $\alpha - \beta$ into several components in such a way that A and B now belong to different components

It is less objectionable in Russian than in English (example 1) to introduce complicated subordinate clauses by the words "of which," often deeply buried in the middle of a clause. In English, it usually helps (see examples 2 and 3) to substitute "whose" for "of which" (the notion that "whose" can only mean "of whom" is out of date), or to use "with" or some similar construction. Example 4 shows a case (not at all rare) where "whose" is impossible and "of which" is clumsy; in these cases a phrase like "in such a way that... " is almost mandatory, and almost a cure-all.

English Index

Russian Index

Transliteration of Cyrillic							
Cyrillic		MR (new)	old (MR, Zbl)	Zbl (new)	ANSI	LC	ISI
А	а	a	a	a	a	a	a
Б	б	b	b	b	b	b	b
В	в	v	v	v	v	v	v
Г	г	g	g	g	g	g, h[1]	g
Д	д	d	d	d	d	d	d
Ђ	ђ[3]	dj	Đ đ	dj	đ	đ	
Ѓ	ѓ[4]	ǵ			ǵ	ǵ	
Е	е	e	e	e	e	e	e
Ё	ё	ë	ë	e	ë	ë	e
Є	є[1,2]	ē	ē	je	ye	i͡e	
Ж	ж	zh	ž	zh	zh, ž[3,4]	zh, z͡h[1], ž[3,4]	zh
З	з	z	z	z	z	z	z
И	и	i	i	i	i	i, y[1]	i
І	і[1,2]	ī	ī	yi	ī	i	
Ї	ї[1]	ï	ï	yi	ï	ï	
Ј	ј[3,4]	j	j	j	j	j	j
Й	й	ĭ	ĭ	j	ĭ	ĭ	i
К	к	k	k	k	k	k	k
Л	л	l	l	l	l	l	l
Љ	љ[3,4]	lj	lj	lj	lj	lj	
М	м	m	m	m	m	m	m
Н	н	n	n	n	n	n	n
Њ	њ[3,4]	nj	nj	nj	nj	nj	
О	о	o	o	o	o	o	o
П	п	p	p	p	p	p	p
Р	р	r	r	r	r	r	r
С	с	s	s	s	s	s	s
Т	т	t	t	t	t	t	t
Ћ	ћ[3]	ć	ć	cj	ć	ć	
Ќ	ќ[4]	ḱ			ḱ	ḱ	
У	у	u	u	u	u	u	u
Ў	ў[2]	ŭ			ŭ	ŭ	
Ф	ф	f	f	f	f	f	f
Х	х	kh	h	kh	kh, h[3,4]	kh, h[3,4]	kh
Ц	ц	ts	c	ts	ts, c[3,4]	t͡s, ts[1], c[3,4]	ts
Ч	ч	ch	č	ch	ch, č[3,4]	ch, č[3,4]	ch
Џ	џ[3,4]	dzh	dž	dzh	dž	dž	
Ш	ш	sh	š	sh	sh, š	sh, š[3,4]	sh
Щ	щ	shch	šč	shch, sht[5]	shch	shch, sht[5]	shch
[Ъ[5]]	ъ	[″] ″	″	”, y[5]	”, ă[5]	″, ŭ[5]	″
Ы	ы	y	y	y	y	y	y
	ь	′	′	’	’	’	’
Э	э	è	è	eh	ė	ė	é
Ю	ю	yu	ju	yu	yu	i͡u	yu
Я	я	ya	ja	ya	ya	i͡a	ya

1 Ukrainian; 2 Belorussian; 3 Serbian; 4 Macedonian; 5 Bulgarian.